FLOWER GUIDE
for HOLIDAY WEEKENDS

in Eastern Canada and Northeastern U.S.A.

This Book Compliments of CANADIAN SCIENCE PUBLISHING | ÉDITIONS SCIENCES CANADA

Publisher of NRC Research Press Journals

Canadian Science Publishing
1200 Montreal Road
Building M-55
Ottawa, ON K1A 0R6
P: 613-990-7873
F: 613-952-7656
pubs@nrcresearchpress.com
www.nrcresearchpress.com

FLOWER GUIDE for HOLIDAY WEEKENDS

in Eastern Canada and Northeastern U.S.A.

Ellen Wynne Larsen
and Betty Ida Roots

NRC Press – Presses du CNRC
An Imprint of NRC Research Press
Ottawa 2005

©2005 National Research Council of Canada

All rights reserved. No part of this publication may be reproduced in a retrieval system, or transmitted by any means, electronic, mechanical, photocopying, recording or otherwise, without the prior written permission of the National Research Council of Canada, Ottawa, Ontario K1A 0R6, Canada.

Printed in Canada on acid-free paper.

ISBN 0-660-19096-6

NRC No. 46327

National Library of Canada cataloguing in publication data

Larsen, Ellen Wynne

Flower guide for holiday weekends in Eastern Canada and Northeastern U.S.A.

Includes bibliographical references.

Issued by the National Research Council Canada.

ISBN 0-660-19096-6

1. Wild flowers – Canada, Eastern – Pictorial works.
2. Wild flowers – Northeastern States – Pictorial works.
3. Wild flowers – Canada, Eastern – Identification.
4. Wild flowers – Northeastern States – Identification.

I. Roots, Betty I. (Betty Ida)

II. National Research Council Canada.

III. Title.

QK115.L34 2005   582.13'0971   C2003-980274-4

Cautionary Note and Disclaimer

Medical self diagnosis and self medication, as noted frequently in this work, are potentially hazardous and are not recommended. Folkloric information is often flawed, and is given here for historical perspective only. References to the medicinal uses of plants are not intended to replace the medical advice of trained health-care professionals. Liability arising directly or indirectly from the use of any of the information provided here is specifically disclaimed.

CONTENTS

Preface vi

Acknowledgements vii

Introduction 1

Photographic Key to the Wildflowers 3

Victoria Day/Memorial Day 11

Baneberry, red; Beardtongue, hairy; Bellwort, large-flowered; Bloodroot; Blue-eyed grass, common; Cohosh, blue; Columbine; False Solomon's seal, star-flowered;Ginger, wild; Hawthorn; Hepatica; Indian paintbrush; Iris, blue flag; Juneberry; Lady's slipper, large yellow; Leek, wild; Marsh marigold; Meadow rue, early; Pin cherry; Polygala, fringed; Prairie smoke; Pussytoes; Spring beauty; Strawberry, barren; Strawberry, common; Striped coral root; Trillium, large flowered; Trillium, purple; Trout lily; Violet, downy yellow

Canada Day/Independence Day 73

Anemone, Canada; Beardtongue, hairy; Bellwort large-flowered; Blue-eyed grass, common; Columbine; Dogbane, spreading; Goat's beard, yellow; Honeysuckle, hairy; Indian paintbrush; Leek, wild; Milkweed, common; Raspberry, purple-flowering; Rose, smooth; Striped coral root; Sweet clover, white

Civic Holiday/Labor Day 105

Baneberry, red; Bergamot; Blue vervain; Boneset; Cardinal flower; Cohosh, blue; Evening primrose; False Solomon's seal, star-flowered; Goat's beard, yellow; Goldenrod; Helleborine; Leek, wild; Milkweed, common; Mint, wild; Ragweed; Raspberry, purple-flowering; Staghorn sumac; Teasel; Thistle, bull

Glossary 144

Bibliography 146

Indexes

   Common names 148
   Scientific names 149

PREFACE

The book is intended as a companion for visitors to the countryside on long holiday weekends who are intrigued by the flowers they see but find field guides formidable. It is expected that the lively and enthusiastic text, as well as the photographs, will act as a gateway for readers to deepen their interest in natural history.

The flowers chosen, with perhaps one exception, occur widely over eastern Canada and the northeastern United States. They do not have exotic habitats and the majority can be spotted without even leaving the car.

The special features of this book which will be useful to the intended audience are: (1) it is organized in sections according to flowering times coinciding with the three North American summer holiday weekends (Victoria Day/Memorial Day; Canada Day/Independence Day; Civic Holiday/Labor Day); (2) flower close-ups, plant identification views, and habitat photographs are included; and (3) pictures of later stages of seed development are shown when these are likely to be seen.

Acknowledgements

A book needs the efforts of more people than just the authors and we are pleased to acknowledge the contributions made to bringing this book to fruition.

We thank Gerald Neville of NRC Research Press for coordinating the book through to completion. We thank Robert Forrest, also of NRC Research Press for his meticulous attention, especially to the layout of the book making it visually appealing. We also thank the designer of the cover for making it so attractive.

We are grateful to the anonymous reviewers for their helpful comments.

Our portraits were taken by Dr. Roberta Bondar and we thank her for generously giving us her time and expertise.

We would like to thank Mary Mulder of Nikon Canada who kindly supplied the Super Coolscan 8000 for creating digital files from all the transparencies.

Last but not least, we greatly appreciate the encouragement of our friends and colleagues through all stages in the development of the book.

Introduction

The idea for this book came after the authors had spent several holiday weekends indulging their shared passion for photography and love of wild flowers. It is intended to be a pocket companion during long holiday weekends, and features flowers which are common in most of eastern Canada and northeastern United States. Many of them can be spotted from a car, moving not too quickly, along country roads and are to be found not too far away from roadsides. Thus, the book is arranged in sections indicating what may be seen on holiday weekends. We hope that these accounts will lure readers to examine flowers at close quarters.

For each of the selected species, photographs are included to enable identification of the plants and flowers and to show habitats. Other notable features such as seed pods are also illustrated. Information on aspects of the biology of the plants, such as pollination, changes in form and /or color with age and opening of flowers with time of day and other factors, is provided. We also include flowering dates. These dates are inclusive for the entire range of each plant. Flowers appear earlier in more southern areas and the flowering period lasts longer. We monitored the appearance of flowers in southern Ontario and noted that, as expected, it was earlier in more southern areas and the flowering period was longer. Even within a given area, flowering dates vary from year to year depending on conditions. We found that this was particularly true of spring flowers. Special habitat requirements which dictate where different species are to be found are described. Usually we have given only one of several possible common names. Because the food and medicinal value of plants has been a subject of fascination for centuries, we have included some of the more interesting uses we have read about. We have relied on secondary sources for this information and provide it for its general interest. We do not recommend that readers ingest any plants on the basis of information given here, not only so that they do not poison themselves but also in the interest of plant conservation.

Taking the photographs was challenging! Many species are low to the ground. This challenge was met by lying on the ground and by using a tripod with special clamps or a Kirk's low pod, often with a focusing rail which obviated the need to move the camera and ourselves. For photographs to be clear and useful, it is necessary to avoid distracting backgrounds. Using limited depths of field so the background was out of focus frequently was not possible or adequate. Eventually we devised a screen of Lexan® pattern material $1/64^{th}$ of an inch thick which we could place behind the plant and which diffused the background. The screen also helped shield from wind.

We used Canon and Minolta 35 mm SLR cameras with a variety of lenses including zooms, macros and screw-in close focusing lenses and of course we always used tripods. Polarizing filters were used to reduce glare. Reflectors were used to direct more light into shady situations. We rarely used flash but when we did so we took care that the resulting photograph reflected the color one would see in daylight. The apparent color of flowers varies even in daylight depending on the quality of light. This may be appreciated by comparing the two photographs of bergamot. In the plant identity photograph, taken in bright light, the flowers are pink whereas in the macro shot taken in shade they are a more bluish mauve.

Other hazards encountered, were poison ivy, not always successfully avoided, mosquitoes and black flies. Repellent, bug nets and surgical gloves were the weapons of choice in the battles against insects.

We almost always photographed together because it was more fun and as most photographers know, the most important piece of equipment is often another pair of hands.

PHOTOGRAPHIC KEY *to the* Wildflowers

Flower Guide for Holiday Weekends

Baneberry, red
Page 12

Beardtongue, hairy
Page 14

Bellwort, large-flowered
Page 16

Bloodroot
Page 18

Blue–eyed grass
Page 20

Cohosh, blue
Page 22

Columbine
Page 24

False Solomon's seal, star-flowered
Page 26

Ginger, wild
Page 28

Hawthorn
Page 30

Hepatica
Page 32

Indian paintbrush
Page 34

Iris, blue flag
Page 36

Juneberry
Page 38

Lady's slipper, large yellow
Page 40

4

Victoria Day / Memorial Day

Leek, wild
Page 42

Marsh marigold
Page 44

Meadow rue, early
Page 46

Pin cherry
Page 48

Polygala, fringed
Page 50

Prairie smoke
Page 52

Pussytoes, smaller
Page 54

Spring beauty
Page 56

Strawberry, barren
Page 58

Strawberry, common
Page 60

Striped coral root
Page 62

Trillium, large-flowered
Page 64

Trillium, purple
Page 66

Trout lily
Page 68

Violet, downy yellow
Page 70

5

Flower Guide for Holiday Weekends

Anemone, Canada
Page 74

Beardtongue, hairy
Page 76

Bellwort, large-flowered
Page 78

Blue–eyed grass
Page 80

Columbine
Page 82

Dogbane, spreading
Page 84

Goat's beard, yellow
Page 86

Honeysuckle, hairy
Page 88

Indian paintbrush
Page 90

Canada Day / Independence Day

Leek, wild
Page 92

Milkweed, common
Page 94

Raspberry, purple-flowering
Page 96

Rose, smooth
Page 98

Striped coral root
Page 100

Sweet clover, white
Page 102

Flower Guide for Holiday Weekends

Baneberry, red
Page 106

Bergamot
Page 108

Blue vervain
Page 110

Boneset
Page 112

Cardinal flower
Page 114

Cohosh, blue
Page 116

Evening primrose
Page 118

False Solomon's seal,
star-flowered
Page 120

Goat's beard, yellow
Page 122

Goldenrod
Page 124

8

Civic Holiday / Labor Day

Helleborine
Page 126

Leek, wild
Page 128

Milkweed, common
Page 130

Mint, wild
Page 132

Ragweed
Page 134

Raspberry, purple-flowering
Page 136

Staghorn sumac
Page 138

Teasel
Page 140

Thistle, bull
Page 142

Victoria Day/Memorial Day

Baneberry, red — *Actaea rubra* (Aiton) Willd. 12
Beardtongue, hairy — *Penstemon hirsutus* (L.) Willd. 14
Bellwort, large-flowered — *Uvularia grandiflora* Sm. 16
Bloodroot — *Sanguinaria canadensis* L. 18
Blue–eyed grass, common — *Sisyrinchium montanum* Greene 20
Cohosh, blue — *Caulophyllum thalictroides* (L.) Michx. 22
Columbine — *Aquilegia canadensis* L. 24
False Solomon's seal, star-flowered — *Maianthemum stellatum* (L.) Link 26
Ginger, wild — *Asarum canadense* L. 28
Hawthorn — *Crataegus* L. spp. 30
Hepatica — *Anemone acutiloba* (DC.) G. Lawson 32
Indian paintbrush — *Castilleja coccinea* (L.) Spreng. 34
Iris, blue flag — *Iris versicolor* L. 36
Juneberry — *Amelanchier* Medik. spp. 38
Lady's slipper, large yellow — *Cypripedium calceolus* L. var. *pubescens* (Willd.) Correll 40
Leek, wild — *Allium tricoccum* Aiton 42
Marsh marigold — *Caltha palustris* L. 44
Meadow rue, early, ♂ & ♀ — *Thalictrum dioicum* L. 46
Pin cherry — *Prunus pennsylvanica* L. f. 48
Polygala, fringed — *Polygala paucifolia* Willd. 50
Prairie smoke — *Geum triflorum* Pursh 52
Pussytoes, smaller — *Antennaria howellii* Greene subsp. *neodioica* (Greene) R.J. Bayer 54
Spring beauty — *Claytonia virginica* L. 56
Strawberry, barren — *Waldsteinia fragarioides* (Michx.) Tratt. 58
Strawberry, common — *Fragaria virginiana* Miller subsp. *virginiana* 60
Striped coral root — *Corallorhiza striata* Lindl. 62
Trillium, large-flowered — *Trillium grandiflorum* (Michx.) Salisb. 64
Trillium, purple — *Trillium erectum* L. 66
Trout lily — *Erythronium americanum* Ker Gawl. subsp. *americanum* 68
Violet, downy yellow — *Viola pubescens* Aiton 70

Flower Guide for Holiday Weekends

Plant Name: Baneberry, red
Actaea rubra (Aiton) Willd.
Flower Size: Cluster diameter is 2.5–3.75 cm (1–1½ in)
Plant Height: 30–80 cm (1 ft–2 ft 8 in)
Habitat: A variety of woodlands
Perennial
Flowering Times: May–June

Notes:

The baneberries look like small bushes and are found scattered in woods. The rounded flower clusters of the red baneberry have a delicate fragrance although this is not mentioned in flower books we have read. The most striking aspect of the red baneberry is the shiny red cluster of berries which appear in late summer. A shaft of sunlight coming through the forest canopy and reflecting off the incredibly shiny berries enlivens many a late summer walk. White baneberry grows in similar areas and is distinguished by its white berries (called 'doll's eyes') on thick flower stalks.

Despite the fact that baneberries have toxic substances in all parts of the plant, aboriginal women apparently used a tea prepared from roots after giving birth. The tea was also drunk as a cure for syphilis and rheumatism.

Ruffed grouse eat the berries with impunity.

12

Victoria Day / Memorial Day

Red Baneberry

Flower Guide for Holiday Weekends

Plant Name: Beardtongue, hairy
Penstemon hirsutus (L.) Willd.
Flower Size: 2.5 cm (1 in) long
Plant Height: 30–90 cm (1–3 ft)
Habitat: Dry, rocky areas, fields
Perennial
Flowering Times: June–July

Notes:

The delicate lavender of this member of the Snapdragon family contrasts with the brilliant scarlet of Indian paint brush and the vivid yellows of members of the daisy family with which they are sometimes associated. The exotic looking trumpet shaped flower with a hairy stem and a sterile, hairy stamen allow identification of this species in an otherwise confusing group. In the inset photograph to the right, petal material has been removed to reveal the four conventional stamens and the yellow brush-like sterile stamen which is about 8 mm ($1/3$ in) long.

Victoria Day / Memorial Day

Hairy Beardtongue

Flower Guide for Holiday Weekends

Plant Name: Bellwort, large-flowered
Uvularia grandiflora Sm.
Flower Size: 2.5–5 cm (1–2 in)
Plant Height: 20–50 cm (8–20 in)
Habitat: Hardwood forests
Perennial
Flowering Times: April–June

Notes:

Our area is blessed with a great number of species of lilies and the Bellwort is a particularly lovely one with its nodding yellow flowers and a stem which seems to go through each leaf. For us, it is a balm to winter weary eyes. It flowers in the forest, early in spring, before the canopy has leafed out. Shortly after the Bellwort leaves have unfurled the green pyramidal berries appear. See Photograph below. Since our specimens cohabit with poison ivy, our pictures were taken with great (but on one occasion, insufficient) care. Apparently, the flower looks like the human uvula at the back of the throat and this, as well as its use as a medicine for sore throat, gave rise to its generic name, *Uvularia*.

Native Americans used it for a variety of other ailments from toothache to rheumatism. Young shoots were also eaten like asparagus.

Large-Flowered Bellwort

16

Victoria Day / Memorial Day

Large-Flowered Bellwort

Flower Guide for Holiday Weekends

Plant Name: Bloodroot
Sanguinaria canadensis L.
Flower Size: 2–5 cm (³/₄–2 in)
Plant Height: 5–15 cm (2–6 in)
Habitat: Rich moist woods, hardwood forests, and roadsides
Perennial
Flowering Times: April–May

Notes:

There are many beautiful and unusual features of this early spring plant. The leaves are an odd shape and they envelop the flower bud until the flower bearing shoot grows above the leaf. After the flowering of this elegant white poppy, the leaves enlarge dwarfing the developing seed pods. Like several other plants, its flowers close at night. We particularly appreciate this plant because in addition to growing in rich, moist woodlands, it is sometimes found in profusion along roadsides otherwise made barren by layers of dirt from winter sanding. At night, one's car lights reflect off the closed flowers, as if they were torches lighting the way to the cottage and a refreshing weekend in the country.

The name, sanguinaria, refers to the red sap of the roots and stems used as a general dye and by native Americans as a war paint and insect repellent. The juice is poisonous in large doses but in minute doses it stimulates the appetite. It contains an alkaloid, sanguinarine, which is used in toothpaste and mouthwashes to inhibit plaque formation.

Victoria Day / Memorial Day

Bloodroot

Plant Name: Blue-eyed grass, common (or Montane) *Sisyrinchium montanum* Greene
Flower Size: 1.3 cm (½ in)
Plant Height: 10–60 cm (4 in–2 ft)
Habitat: In relatively moist areas, meadows, prairies, ditches, and shores
Perennial
Flowering Times: June–July

Notes:

A plant often seen in meadows during the summer, it's grass-like narrow leaves and small blue to purple flower seem unlike the showier members of the iris family to which it belongs. Common blue-eyed grass may be distinguished from Pointed blue-eyed grass (*S.angustifolium*) by the fact that it has a short-stalked flower whereas that of the latter is long-stalked. When photographing in late afternoon one discovers that the flowers begin to close, so, as with other flowers like blood root, goat's beard and trout lily, it is necessary to plan photography around their timetables.

Native Americans used several species as laxatives. The corms of a related European species were claimed, in 1633, to be eaten "to procure lust and lecherie".

Victoria Day / Memorial Day

Common Blue-Eyed Grass

Flower Guide for Holiday Weekends

Plant Name: Cohosh, blue
Caulophyllum thalictroides (L.) Michx.
Flower Size: 3–6 cm (1¼–2½ in)
Plant Height: 30–80 cm (1 ft–2 ft 8 in)
Habitat: Moist woods and hardwood forests, particularly in limestone areas
Perennial
Flowering Times: April–May
Notes:

In early spring, the furled leaves of this woodland plant look so purple and strange that even the botanically challenged are moved to find out what it is! Three successive stages in leaf development are shown, mid and bottom right and below.

We always enjoy the development of the ungainly leaves into more aesthetically pleasing foliage. At this stage, our photographic efforts are observed with great interest by black flies. Although the flowers are not as vivid as the large dark blue, berry-like seeds which will develop later in the season, the flowers ensure that the seed is the product of cross-fertilization by maturing female and male parts of the flower at different times.

The roots were used by Native Americans to facilitate childbirth and it was listed in the US Pharmacopoeia until 1905. Latter day herbalists still make medicinal claims for treating everything from worms to gout but admit that patients with hypertension or heart disease should give this curative a miss.

Victoria Day / Memorial Day

Blue Cohosh

Flower Guide for Holiday Weekends

Plant Name: Columbine
                Aquilegia canadensis L.
Flower Size: 3–4 cm (1¼–1½ in) across and up to 4 cm (1½ in) long
Plant Height: 20–90 cm (8 in–3 ft) tall
Habitat: Dry, rocky open areas or in forest edges and clearings
Perennial
Flowering Times: April–July

Notes:

Ontario's native columbine is reddish orange and therefore attractive to hummingbirds which can hover at the flower, insert long tongues into the flower spurs and obtain the nectar at the far tip. Some moths also have long enough tongues to secure nectar and both bird and moth can transfer pollen, which is dangling near the entrance to the spurs, from flower to flower, in some sense "paying" for the nectar. If you see a small hole near the nectary, this may have been made by a short-tongued bee which found a shortcut to obtaining the sugary nectar.

An interesting feature of this flower is that it nods on a curved stem. After fertilization when seeds are forming, the stem becomes erect, (as shown in the photograph on this page). This feature is also seen in the prairie smoke, another flower in this section. Because of the long flowering time and the sequential development of flowers both erect seeds and the nodding flowers are sometimes found on the same plant.

One of the more interesting Native American medicinal uses was that the seeds were rubbed into hair to control lice. Crushed seeds have been used as a perfume.

Columbine

24

Victoria Day / Memorial Day

Columbine

Flower Guide for Holiday Weekends

Star-Flowered False Solomon's Seal

Plant Name: False Solomon's seal, star-flowered *Maianthemum stellatum* (L.) Link (formerly *Smilacina stellata* Desf.)
Flower Size: 6 mm (¼ in) long
Plant Height: 20–50 cm (8–20 in)
Habitat: Prefers moist soil and is found along roadsides and in damp woods, but also grows on sand dunes
Perennial
Flowering Times: May–June

Notes:

This member of the lily family has four or more alternate narrow leaves clasping the stem. Masses of delicate creamy-white flowers at the ends of stems distinguish false Solomon's seals from the true Solomon's seals in which hanging bell-shaped flowers arise from the leaf axils. The green berries go through an attractive phase of being striped with red and blackish red before turning completely red and then almost black. Under the right conditions, dense patches of these plants adorn roadsides producing a subtle lily-like fragrance which wafts upward to an appreciative observer (or photographer).

Victoria Day / Memorial Day

Star-Flowered False Solomon's Seal

Flower Guide for Holiday Weekends

Wild Ginger

Plant Name: Ginger, wild
Asarum canadense L.
Flower Size: 2–4 cm (³/₄–1¹/₂ in)
Plant Height: 15–30 cm (6–12 in)
Habitat: Rich woods with calcareous soils, sometimes moist
Perennial
Flowering Times: April–May

Notes:

The thrill of discovering a tropical-looking flower under the heart-shaped leaves of wild ginger is still vivid after several decades. This usually hidden treasure is exposed briefly while the new leaves are growing (see photographs on right). Sometimes, large patches of ground in shady areas are covered with wild ginger as seen in the habitat photograph below. Usually such a patch is created by vegetative reproduction since new plants grow from tips of underground stems.

The root has been used as a ginger substitute after drying or being candied with sugar, however, as Native American women used it for birth control, we do not recommend culinary experimentation until its constituents are better known!

28

Victoria Day / Memorial Day

Wild Ginger

Plant Name: Hawthorn
Crataegus L. spp.
Flower Size: 1–2 cm (³⁄₈–³⁄₄ in)
Plant Height: up to 12 m (36 ft)
Habitat: Disturbed areas, abandoned fields, in forest openings. Particularly on calcium rich soils
Perennial
Flowering Times: May–June
Notes:

It is easy to identify a tree as a hawthorn but difficult to determine the species. There are many hybrids in this member of the rose family. The name comes from haw meaning fence, and thorn. It was much used in hedging in Europe, and the fruit is also known as a haw. It has some of the longest thorns one is likely to encounter.

In a year of prolific flowering, the trees can be a visual delight when covered with blooms or burdened with innumerable red fruits. Even without flowers, the Hawthorn shrub or tree enlivens an abandoned field with its interesting twisting branches and the bird life it attracts. Loggerhead shrikes store their prey, insects, rodents and even small birds, by impaling them on the large thorns. Poisonous insects are eaten only after two or three days storage by which time the poisons have degraded. People have been known to use them as toothpicks.

Aboriginals used flowers and roots in medicinal teas for heart ailments, and the fruits as winter food. Modern research has found compounds in hawthorns useful as diuretics and for dilating coronary arteries. There is some evidence from animal research that hawthorns contain substances which lower blood cholesterol levels. The wood is suitable for carving.

Victoria Day / Memorial Day

Hawthorn

Flower Guide for Holiday Weekends

Plant Name: Hepatica
Anemone acutiloba (DC.) G. Lawson (formerly *Hepatica acutiloba* DC.)
Flower Size: 1.3–2.5 cm (½–1 in)
Plant Height: 10–15 cm (4–6 in)
Habitat: Deciduous woodlands
Perennial
Flowering Times: April–May

Notes:

One of the earliest spring flowers, Hepaticas bloom along woodland paths in the company of others such as trout lillies, as seen in the lower photograph on the right.The flowers emerge amid last year's leaves which are usually concealed beneath leaf litter. New leaves come up later, see photograph below. The color of the petals varies from blue to pink or white. The name hepatica is derived from the Greek word for liver, *hepar*. In medieval times it was believed that the physical characteristics of plants indicated the medicinal uses of the plant. Because its three-lobed leaves resembled the three lobes of our livers, potions prepared from hepatica were believed to cure liver ailments. This belief persists even today in some places. In north America during the late 19th century the demand for the leaves was so great that, according to one source, more than 450,000 lbs. (204,000 kg) were imported from Europe in a single year. This represents a staggering amount of the small leaves!

The species shown here has pointed leaf lobes (hence *acutiloba)*. A similar species *A. americana,* has rounded lobes.

32

Victoria Day / Memorial Day

Hepatica

Flower Guide for Holiday Weekends

Plant Name: Indian Paintbrush
Castilleja coccinea (L.) Spreng.
Flower Size: 2.5 cm (1 in) but hidden
Plant Height: 30–60 cm (1–2 ft)
Habitat: Meadows, often damp soil
Annual or Biennial
Flowering Times: May–July

Notes:

The scarlet of Indian Paintbrush is visible at great distances in meadows, often to startling effect. The color is actually associated with leaf-like bracts that almost conceal the inconspicuous flowers. The mingling of Indian Paintbrush with buttercups in a wet meadow often looks from a distance like a French impressionist painting. Because the bracts last longer than flowers, paintbrushes appear to "bloom" with successive other species as the season progresses. Hovering insects or birds are needed for pollination as there are no perches available for stationary pollinators. It is a partial root parasite. Vigorous growth of this plant requires both its own photosynthesis and nutrients obtained through a parasitic connection to the roots of another plant. A wide variety of plants serve as hosts. Native Americans used tea made from the flowers for rheumatism and women's ailments as well as a love potion and as a poison!

Victoria Day / Memorial Day

Indian Paintbrush

Flower Guide for Holiday Weekends

Plant Name: Iris, blue flag
Iris versicolor L.
Flower Size: 6.3–10 cm (2½–4 in) in diameter. Sepals 2 cm (¾ in) wide and up to 7.5 cm (3 in) long. Petals up to 2 cm (¾ in) long.
Plant Height: 60–90 cm (2–3 ft)
Habitat: Wetlands, sunlight or light shade, swamps, marshes, wet shores.
Perennial
Flowering Times: May–August

Notes:

The wild irises of moist environments appear to the city dweller like escapees from a garden! Their large, complicated, blue flower seems out of place among the sedges and rushes amongst which they are often found. Indeed the name blue flag is said to derive from Middle English and means "blue reed". This species is a close relative of blue-eyed grass. The three downwardly curving blue and yellow petal-like structures are actually sepals, the true petals point upward. The three styles are also petal-like and curve over the sepals, each concealing an anther. As a pollinating bee or fly pushes its way in to reach the nectar its back brushes first against the receptive area of the style at its tip, depositing pollen acquired from a previously visited flower, and then against the anther gathering fresh pollen to take to another flower.

As beautiful as irises are, their underground storage organs (rhizomes) are poisonous. Apparently, Native North Americans and European colonists used the dried rhizome in small quantities as a diuretic and cathartic. Native Americans also made a poultice which they used on sores and bruises.

Blue Flag Iris

Victoria Day / Memorial Day

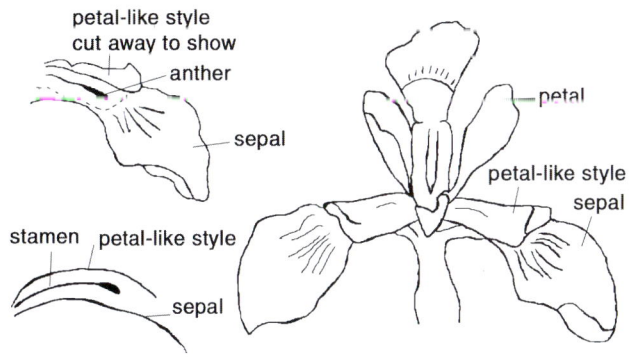

Blue Flag Iris

Plant Name: Juneberry
Amelanchier Medik. spp.
Flower Size: 3 cm (1¾ in)
Plant Height: Up to 12 m (36 ft)
Habitat: Forested and open areas, usually dry
Perennial
Flowering Times: April–June

Notes:

This tree-sized member of the rose family belongs to the large and confusing Serviceberry group of shrubs and small trees. These botanical frustrations never detract from the pleasure of seeing the leafless tree covered in blooms as one of our earliest flowering trees. It can often be spotted in fields from the road although it also adorns woodlands associated with cherry and other deciduous trees.

The berries (which usually mature later than June) serve as food for some birds and mammals. Although humans find June berries tasteless, serviceberries, in general, are said to be high in copper and iron and may be eaten raw, dried, in jams and jellies and as an important constituent of the Native North American travel food, pemmican. In addition to the berries, virtually every part of the tree above ground has been used by wildlife for food. In the past, the wood was used for making pipes and arrows.

Victoria Day / Memorial Day

Juneberry

Plant Name: Lady's slipper, large yellow
Cypripedium calceolus L. var. *pubescens* (Willd.) Correll
Flower Size: The sac-like lower lip is about 5 cm (2 in) long
Plant Height: 10–70 cm (4–28 in)
Habitat: Rich moist woods, bogs
Perennial
Flowering Times: June–August

Notes:

This conspicuous orchid, whose name comes from the Latin for shoe, may be found along shaded roadsides or in a marsh surrounded by horsetails and other plants of swampy areas. Patches of these plants with numerous flowers gently swaying in the breeze are an aesthetic treat long savored. This plant prefers areas rich in limestone and therefore somewhat alkaline soils compared to the pink Lady's Slipper which tolerates more acidic soils.

The leaves and flowers are eaten by white-tailed deer. Native North Americans and 19th century colonists used it as a sedative and extensive harvesting depleted the populations of this plant.

Victoria Day / Memorial Day

Large Yellow Lady's Slipper

Plant Name: Leek, wild
Allium tricoccum Aiton
Flower Size: 5–7 mm (¼ in), umbel 3.8 cm (1½ in) in diameter
Plant Height: 10–30 cm (4–12 in)
Habitat: Rich moist woods, hardwood forests often in limestone areas
Perennial or Biennial
Flowering Times: June–July
Notes:

One of the hopeful signs of spring is the emergence of the lush, green wild leek leaves carpeting hardwood forest floors. This early appearance of the leaves enables the plants to capture all the energy they need, both for underground storage and flower production, before sunlight is prevented from reaching the forest floor by the leafing out of the forest trees. The leek leaves die back very quickly once the sunlight is blocked. The young plants, both above and below ground portions, have a flavorful onion taste and collecting these and the young leaves has become a spring ritual in the Appalachian mountains of the south eastern United States.

Wild Leek

Victoria Day / Memorial Day

Wild Leek

Plant Name: Marsh marigold
Caltha palustris L.
Flower Size: 2.5–4 cm (1–1½ in)
Plant Height: 15–80 cm (6 in–1 ft 8 in)
Habitat: Wet and marshy places, shallow water
Perennial
Flowering Times: April–June

Notes:

Another buttercup lacking true petals, the marsh marigold enlivens many a marsh, stream side and roadside ditch with its luxuriant leaves and intensely yellow flowers. Thoreau called it "A flower-fire bursting up....". In addition to its visual delights marsh marigold has also been used as a pot herb since its leaves were available before northern gardens produced greens. It is not, however, to be eaten raw. In the 18th century, some felt that cows produced yellow butter in spring after eating this plant but a less credulous observer of the time pointed out that cows who ate it generally sickened and died.

Leaf tea mixed with maple syrup was enjoyed by colonists. The flowers, pickled in vinegar were used as a substitute for capers. They were boiled first to remove toxins. It is interesting to note that the sap has been used as a cure for warts.

Victoria Day / Memorial Day

Marsh Marigold

Flower Guide for Holiday Weekends

Plant Name: Meadow rue, early
Thalictrum dioicum L.
Flower Size: 6 mm (¼ in) long
Plant Height: 20–75 cm (8 in–2½ ft)
Habitat: Moist woods
Perennial
Flowering Times: April–May
Notes:

This is yet another member of the buttercup family but it has the distinction of bearing female and male flowers on different plants. How is this of benefit to the species? How is sex determined? What sort of ratio of male to female plants is necessary? These are questions which come to our minds. The male flowers (photograph on far right), with their long white stamens which later become purple, are pendulous and sway with the slightest breeze. This may enhance pollen dispersal but it is no help at all to the photographer trying to get a clear image of these delicate flowers. The female flowers may be distinguished from the male flowers by their long purplish pistils. Neither male nor female flowers have true petals.

Infusions of the roots were used as a purgative and diuretic.

Early Meadow Rue

Victoria Day / Memorial Day

Early Meadow Rue

Flower Guide for Holiday Weekends

Plant Name: Pin cherry
Prunus pennsylvanica L. f.
Flower Size: Small 1.2–1.5 cm (≈½ in) wide
Plant Height: Up to 12 m (36 ft) tall
Habitat: Dry to moist in mixed wood stands, pine and intolerant hardwood stands; forest openings; edges and roadsides
Perennial
Flowering Times: May to early June
Notes:

The flowers of this shrub or small tree appear when the leaves are half grown. They are white and occur as flat topped clusters usually arising from leaf axils. The bright red round fruits, 5–7 mm (¼ in) in diameter, are the earliest of the cherries and are found from August to September. This is a favorite tree for eastern tent caterpillars. Their overwintering egg clusters are glued to branches of the tree and in spring the caterpillars feast on the young leaves often denuding whole plants some of which will have the energy to regrow the leaves later in the spring. The caterpillars have very interesting behavior in terms of weaving "tents" and following each other during foraging. They also have rhythms of foraging and returning to the "tent".

The flesh of the fruits is edible and provides food for mammals and birds. The kernel, however, contains prunasin, a cyanide-like glucoside which becomes hydrogen cyanide during digestion and thus is highly poisonous.

The inner bark was used by aboriginal people to make a tea taken for coughs and internal ailments. Crushed roots were used as a treatment for stomach trouble.

Pin Cherry

Victoria Day / Memorial Day

Pin Cherry

Fringed Polygala

Plant Name: Polygala, fringed
Polygala paucifolia Willd.
Flower Size: 19 mm (³⁄₄ in) long
Plant Height: 7.6–17.8 cm (3–7 in) tall
Habitat: A variety of habitats from rich moist coniferous woods to dry to moist sandy to clayey pine stands and conifer swamps
Perennial
Flowering Times: May–June
Notes:

From a photographer's point of view, these lovely pink flowers may be classified as "belly plants" because of the position assumed to photograph these low down plants. We have not confirmed a report that in addition to its above ground flowers it also has underground self-fertilising flowers. The plants sometimes grow on banks along roadsides under trees. Although the flowers are quite small, we were able to spot some from a slowly moving car! Milkworts such as Polygala are thought to increase milk production when eaten by nursing mothers or cows but in a rare disclaimer, an herbal book admitted this is unfounded. The plant is said to have a mild wintergreen flavor.

Seed dispersal is very interesting in this species. Ants are attracted to the oily liquid in the transparent sac attached to each seed. The ants cut the seeds from ripe fruits and carry them to their nests which may be as far as 10 m (33 ft) away. They then eat the oil filled sac and discard the seed which is undamaged.

Victoria Day / Memorial Day

Fringed Polygala

Plant Name: Prairie smoke
Geum triflorum Pursh
Flower Size: Up to 2 cm (¾ in) long
Plant Height: 15–40 cm (6–16 in)
Habitat: Found only on alvars (limestone plains) where, they are abundant
Perennial
Flowering Times: Late April–July

Notes:

The brownish pink flowers are often found in groups of three, hence the name *triflorum*. They arise on a leafless or nearly leafless stalk with fern-like, evergreen leaves at its base. The fruits bear long 5 cm (2 in), feathery pinkish-gray hairs, the styles, which seen *en masse* give the appearance of smoke wafting across the field. A really impressive sight. As in columbines the flowers droop but after fertilization, when seeds are forming, the stem becomes erect.

Native Americans made a tonic tea from the rhizomes, which was used also as a body wash to ease aches and pains in the sweat-houses. Ripe seeds yield a perfume when crushed.

Victoria Day / Memorial Day

Prairie Smoke

Plant Name: Pussytoes, smaller
Antennaria howellii Greene subsp. *neodioica* (Greene) R.J. Bayer
Flower Size: Heads up to 1.3 cm (½ in) wide
Plant Height: 10–45 cm (4–18 in)
Habitat: Pastures, open woods and lawns
Perennial
Flowering Times: May–July

Notes:

Clusters of fuzzy white flower heads resemble a cat's paw and give this plant its name. Each flower head is made up of disc florets enclosed by greenish or brownish bracts with white tips. Male and female flowers are found on different plants but male plants are rare. Plants grow in open areas in dense groups forming patches of white when in flower. Set off by nearby greenery they are a welcome visual addition to fields and are sought after for dried flower bouquets. Wild turkeys, white-tailed deer and cotton tails evidently find pussytoes aesthetically pleasing from a gastronomic perspective.

Native North Americans had multiple uses for pussytoes; to give themselves and their horses strength and energy, as an anti-pertussive and a vermifuge.

Victoria Day / Memorial Day

Smaller Pussytoes

Plant Name: Spring beauty
Claytonia virginica L.
Flower Size: 1.3–2 cm (½–¾ in) wide
Plant Height: 15–30 cm (6–12 in)
Habitat: In moist woods and thickets. Also in clearings
Perennial
Flowering Times: March–May

Notes:

Spectacular patches of this plant with its clusters of striped flowers (dark pink stripes on a pale pink or white background), are a delight to encounter in spring before trees and shrubs have leafed out. Their small size makes photography a challenge for those of us with mature joints. The flowers close not only at night but also during storms, presumably triggered by low light intensity.

The tubers are said to have a sweet, chestnut-like flavor, and were prized as food by Native Americans and early colonists. The seeds are eaten by wildlife.

Victoria Day / Memorial Day

Spring Beauty

Plant Name: Strawberry, barren
Waldsteinia fragarioides (Michx.) Tratt.
Flower Size: 1.3 cm (½ in) wide
Plant Height: 7–20 cm (3–8 in)
Habitat: Woods, thickets, shaded hillsides and clearings in dry to moist soil
Perennial
Flowering Times: April–June

Notes:

Barren strawberries are a cheerful sight on a stroll through a spring wood. The tripartite leaves and general appearance resemble those of the common strawberry but this plant has yellow flowers and produces only a dry fruit with one seed not a succulent berry, hence barren. Even without their flowers, it is not too hard to distinguish this plant from the common strawberry. The teeth on the leaflets appear more irregular and the leaflets are more wedge shaped than the oval strawberry leaves. Barren strawberry also tolerates deeper shade than the common strawberry. Although these plants are usually found in patches (see lower right photograph) they lack the runners so important to common strawberry propagation.

Victoria Day / Memorial Day

Barren Strawberry

Flower Guide for Holiday Weekends

Plant Name: Strawberry, common
Fragaria virginiana Miller subsp. *virginiana*
Flower Size: 1.5–2.5 cm (⁵/₈–1 in) wide in clusters of 2–15 on stalks 7.5–15 cm (3–6 in) tall.
Plant Height: Creeper, 7.5–15 cm (3–6 in)
Habitat: Open fields, roadsides, edges of woods and in disturbed areas
Perennial
Flowering Times: May–June
Notes:

This common low-growing plant has tripartite leaves whose stalks are longer than those bearing flowers. The fruits, although small, 0.5–2 cm (¹/₄–³/₄ in) in diameter, are delicious and well worth taking the trouble to pick. When photographing edible berries we restrain ourselves until after the "shot" when we feel a special reward is in order. They provide food for a variety of non-human animals some of which eat their leaves and flowers as well as their berries. Strawberries often propagate themselves using runners. It is interesting to see that plants attached by runners will have large dark green leaves in a shaded area, and smaller, sometimes reddish colored leaves, if in open sun. This is just one of many examples of the plasticity of plants in adapting to different environments.

Fresh leaf tea is said to be good for sore throats. Leaf and root teas were used by Native Americans as a nerve tonic and for ailments of almost all internal organs.

Common Strawberry

Victoria Day / Memorial Day

Common Strawberry

Plant Name: Striped coral root
Corallorhiza striata Lindl.
Flower Size: 8–20 mm (³⁄₈–³⁄₄ in) long. Raceme with 3–30 flowers
Plant Height: 20–50 cm (8–20 in)
Habitat: Rich moist woods with calcareous or alkaline soil
Perennial
Flowering Times: May–August

Notes:

This strange plant, without a sign of green, is an orchid which lacks chlorophyll. Unlike green plants which can make sugars using photosynthesis, the striped coral root must gain its nutrition by other means. On the basis of what is known for the spotted coral root (*C. maculata*) it is thought that the striped coral root obtains carbon compounds through association with fungi (mycorrhizae). It is hard to imagine that it is fairly closely related to the Lady's Slippers. The name coral root comes from the Greek words for coral and root (*rhiza*), but the coral-like "roots" are in fact rhizomes. The pinkish-yellow or white flowers with conspicuous reddish purple stripes also conjure up visions of coral. Considering that most plants were found useful by aboriginal peoples, it is not surprising to learn that the root was used as a sedative and to reduce fevers.

Victoria Day / Memorial Day

Striped Coral Root

Large-Flowered Trillium

Plant Name: Trillium, large-flowered
Trillium grandiflorum (Michx.) Salisb.
Flower Size: 5–10 cm (2–4 in) wide
Plant Height: 20–45 cm (8–18 in)
Habitat: Rich woods and thickets, usually on basic or neutral soils
Perennial
Flowering Times: April–June

Notes:

Masses of these beautiful white flowers carpet the woods in spring. As the flowers age they turn pinkish before they wither and die. Sometimes flowers may be seen with green stripes or which are entirely green. These color variations are caused by infection of the phloem (nutrient conducting vessels) by a mycoplasma-like organism. This infection may also cause enlarged sepals, distorted stamens, and aborted pistils. Other oddities such as multiple petals are rare but we have followed one such plant which produces on the order of 30 petals but no pistils or stamens. It has produced flowers regularly over 10 years. Nearby trilliums appear to be perfectly normal.

White tailed deer eat both leaves and flowers. The Menomini made extensive use of the roots for a variety of ailments. A poultice of grated raw root was used to reduce swelling of the eye, and teas were drunk for cramps and menstrual irregularities.

Victoria Day / Memorial Day

Large-Flowered Trillium

Plant Name: Trillium, purple
Trillium erectum L.
Flower Size: 6.3 cm (2½ in) wide
Plant Height: 20–40 cm (8–16 in)
Habitat: Rich woods. Moist to dry
Perennial
Flowering Times: April–June

Notes:

Not nearly as plentiful as the white trillium, small clumps of this bronzy maroon (rather than purple) flowered plant are found in similar habitats. The leaves of this plant are a dark green whereas those of the white trillium are a light green. The flowers have an unpleasant odor which together with the color attracts carrion flies which serve as pollinators. In the past this same unpleasant odor also led to the plant being used to treat gangrene on the premise that remedies have properties in common with the disease.

White tailed deer however are not deterred and eat flowers and leaves of this as well as the white trillium.

Also known as bethroot (birth root), a tea prepared from the rootstock (rhizome) was used by Native Americans to induce childbirth and help in labor. It was used also for coughs, bowel troubles, and for menstrual disorders and menopause as well as an aphrodisiac!

It is interesting to note that the rhizome contains steroids which may be the basis for its therapeutic properties.

Victoria Day / Memorial Day

Purple Trillium

Flower Guide for Holiday Weekends

Plant Name: Trout lily
Erythronium americanum Ker Gawl. subsp. *americanum*
Flower Size: 18–40 mm (¾–1½ in) long; 2.5 cm (1 in) wide
Plant Height: 10–30 cm (4–12 in)
Habitat: In moist woods and thickets
Perennial
Flowering Times: March–June

Notes:

It is a delight to see these dainty nodding yellow flowers brightening the woods. They arise from corms. Usually a few two-leaved fertile plants with flowers are found amongst colonies of single leaved sterile shoots, offshoots from the corm bases. Where conditions are sufficiently sunny, dense patches of flowers may be seen. It is said that trout lilies are pollinated mostly by ants and that seeds may take as long as seven years to produce a flowering plant.

The plant is variously known as trout lily, so called because the markings on the mottled leaves resemble those of brook trout; adder's tongue lily because of the purplish points of young leaves emerging in spring; and dogtooth violet, a name which originates from the tooth like shape of the white corm. This latter name is inappropriate since the plant is not a violet but a lily.

Iroquois women ate raw leaves as a contraceptive. Other uses of leaves and roots were as teas for fevers and as poultices to draw out splinters and reduce swelling. Water extracts of the plant have been shown to have anti-bacterial properties.

68

Victoria Day / Memorial Day

Trout lily

Flower Guide for Holiday Weekends

Plant Name: Violet, downy yellow
Viola pubescens Aiton
Flower Size: 2 cm (¾ in) wide
Plant Height: 10–45 cm (4–18 in)
Habitat: Varied, moist to dry, clayey to sandy, rich woods, hardwood swamps
Perennial
Flowering Times: May–June

Notes:

Obviously, not all violets are violet. Their lovely colors and interesting petal shapes combine to make much loved spring blooms. The heart-shaped leaves are 5–12.5 cm (2–5 in) wide. The plant is soft and hairy with solitary yellow flowers having maroon veins in the lower three petals; the lower two petals form a short spur while the two lateral ones are bearded.

The flowers, leaves, roots and seeds were all used in medicines by aborigines. Rabbits, mice, and various birds such as woodcocks and wild turkey eat the leaves, flowers, and seeds.

Downy Yellow Violet

Victoria Day / Memorial Day

Downy Yellow Violet

Canada Day/Independence Day

Anemone, Canada — *Anemone canadensis* L. 74
Beardtongue, hairy – *Penstemon hirsutus* (L.) Willd. 76
Bellwort, large flowered— *Uvularia grandiflora* Sm. 78
Blue-eyed grass, common — *Sisyrinchium montanum* Greene 80
Columbine — *Aquilegia canadensis* L. 82
Dogbane, spreading — *Apocynum androsaemifolium* L. 84
Goat's beard, yellow — *Tragopogon dubius* Scop. 86
Honeysuckle, hairy — *Lonicera hirsuta* Eaton 88
Indian paintbrush — *Castilleja coccinea* (L.) Spreng. 90
Leek, wild — *Allium tricoccum* Aiton 92
Milkweed, common — *Asclepias syriaca* L. 94
Raspberry, purple-flowering — *Rubus odoratus* L. 96
Rose, smooth — *Rosa blanda* Aiton 98
Striped coral root — *Corallorhiza striata* Lindl. 100
Sweet clover, white — *Melilotus alba* Medik. 102

Plant Name: Anemone, Canada
 Anemone canadensis L.
Flower Size: 2–4 cm (³/₄–1¹/₂ in) in diameter. Sepal length 1–2.5 cm (³/₈–1 in)
Plant Height: 20–70 cm (8 in–2 ft 4 in)
Habitat: Along shores and dry, open areas and along roadsides
Perennial
Flowering Times: June–July

Notes:

The origin of the hard to pronounce name, Anemone, is in dispute. Some say it derives from the Greek word for wind and others from the Semitic name for the God, Adonis, from whose blood the flower is said to have sprung. We are stuck with the name, even if it brings to mind neither wind nor gods but small marine creatures with many tentacles.

The anemones are characterized by deeply cut leaves and stems with single blossoms. The Canada anemone is further distinguished by having leaves with no stems. What look like large white petals are really sepals which also protect the rest of the flower when it is not open. It is likely that the dense patches of these plants are the result of vegetative reproductions from roots or rhizomes as well as sexual reproduction mediated by flowers. Our favorite patch is in a moist, shady thicket, adjacent to woods although they are found in drier habitats as well. As with other plants which close up under conditions where pollinators are scarce, they have to be photographed whenever they are ready. Once pollinated the flowers remain closed.

Ironically, despite the exquisite beauty of the flower, anemones contain toxic substances. Nevertheless aboriginal peoples chewed roots to clear the throat before singing and used concoctions of roots and leaves to wash sores.

Canada Day / Independence Day

Canada Anemone

75

Flower Guide for Holiday Weekends

Plant Name: Beardtongue, hairy
Penstemon hirsutus (L.) Willd.
Flower Size: 2.5 cm (1 in) long
Plant Height: 30–90 cm (1–3 ft)
Habitat: Dry, rocky areas, fields
Perennial
Flowering Times: June–July

Notes:

The delicate lavender of this member of the Snapdragon family contrasts with the brilliant scarlet of Indian paint brush and the vivid yellows of members of the daisy family with which they are sometimes associated. The exotic looking trumpet shaped flower with a hairy stem and a sterile, hairy stamen allow identification of this species in an otherwise confusing group. In the inset photograph to the right, petal material has been removed to reveal the four conventional stamens and the yellow brush-like sterile stamen which is about 8 mm ($1/3$ in) long.

Hairy Beardtongue

Canada Day / Independence Day

Hairy Beardtongue

Plant Name: Bellwort, large-flowered
Uvularia grandiflora Sm.
Flower Size: 2.5–5 cm (1–2 in)
Plant Height: 20–50 cm (8–20 in)
Habitat: Hardwood forests
Perennial
Flowering Times: April–June

Notes:

Our area is blessed with a great number of species of lilies and the Bellwort is a particularly lovely one with its nodding yellow flowers and a stem which seems to go through each leaf. The stem zig-zagging through the bottom of each leaf is easily seen in the lower right photograph. Shortly after the leaves have unfurled, the pyramidal berries appear (upper right photograph) and they continue to develop during the rest of the summer. Since our specimens cohabit with poison ivy, our pictures were taken with great (but on one occasion, insufficient) care. The Bellwort's generic name, (*Uvularia*), derives from its supposed resemblance to the human uvula at the back of the throat which also led to its use as a medicine for sore throat.

Native Americans used it for a variety of other ailments from toothache to rheumatism. Young shoots were also eaten like asparagus.

Canada Day / Independence Day

Large-Flowered Bellwort

Plant Name: Blue-eyed grass, common (or Montane) *Sisyrinchium montanum* Greene
Flower Size: 1.3 cm (½ in)
Plant Height: 10–60 cm (4 in–2 ft)
Habitat: In relatively moist areas, meadows, prairies, ditches and shores
Perennial
Flowering Times: June–July

Notes:

A plant often seen in meadows during the summer, it's grass-like narrow leaves and small blue to purple flower seem unlike the showier members of the iris family to which it belongs. Common blue-eyed grass may be distinguished from Pointed blue-eyed grass (*S. angustifolium*) by the fact that it has a short-stalked flower whereas that of the latter is long-stalked. When photographing in late afternoon one discovers that the flowers begin to close, so, as with other flowers like blood root, goat's beard and trout lily, it is necessary to plan photography around their timetables.

Native Americans used several species as laxatives. The corms of a related European species were claimed, in 1633, to be eaten "to procure lust and lecherie".

Canada Day / Independence Day

Blue-Eyed Grass

Flower Guide for Holiday Weekends

Plant Name: Columbine
Aquilegia canadensis L.
Flower Size: 3–4 cm (1¼–1½ in) across and up to 4 cm (1½ in) long
Plant Height: 20–90 cm (8 in–3 ft) tall
Habitat: Dry, rocky open areas or in forest edges and clearings
Perennial
Flowering Times: April–July

Notes:

Ontario's native columbine is reddish orange and therefore attractive to hummingbirds which can hover at the flower, insert long tongues into the flower spurs and obtain the nectar at the far tip. Some moths also have long enough tongues to secure nectar and both bird and moth can transfer pollen, which is dangling near the entrance to the spurs, from flower to flower, in some sense "paying" for the nectar. If you see a small hole near the nectary, this may have been made by a short-tongued bee which found a shortcut to obtaining the sugary nectar.

An interesting feature of this flower is that it nods on a curved stem, seen in both photographs. After fertilization when seeds are forming, the stem becomes erect, as seen in the upper photograph. This feature is also seen in the prairie smoke, see Victoria Day / Memorial Day section. In the columbine seed head there are five small tubes containing small dark seeds which are shaken out by wind. Because of the long flowering time and the sequential development of flowers both erect seeds and the nodding flowers are sometimes found on the same plant.

One of the more interesting Native American medicinal uses was that the seeds were rubbed into hair to control lice. Crushed seeds have been used as a perfume.

Canada Day / Independence Day

Columbine

Flower Guide for Holiday Weekends

Plant Name: Dogbane, spreading
Apocynum androsaemifolium L.
Flower Size: 8 mm (1/3 in) wide
Plant Height: 33–132 cm (1–4 ft)
Habitat: Roadsides, at edges of dry forests, in fields and thickets
Perennial
Flowering Times: June–July
Notes:

The flowers, drooping pink bells with rose stripes inside, occur mostly in showy terminal clusters but may also arise from the axils of leaves. The plant has a milky latex.

The graceful bell-like flowers and interesting pods found on this plant give no indication that potentially poisonous glycosides lurk within. The common and generic names leave no doubt that people discovered this danger early (*Apocynum* is Greek for "away dog". Why dogs were singled out isn't clear since cattle (which after all are more likely to eat plants) are said to receive heart muscle damage from eating this plant. Even insects have trouble with the innocent looking flowers if their mouth parts get caught between the stamens. As for people, the plant was used by some aboriginals for making thread and bowstrings while others used roots for remedies for everything from syphilis to nervousness. It is sometimes known as wild ipecac as it contains this emetic.

Spreading Dogbane

Canada Day / Independence Day

Spreading Dogbane

Flower Guide for Holiday Weekends

Plant Name: Goat's beard, yellow
 Tragopogon dubius Scop.
Flower Size: 2.5–6.3 cm (1–2½ in)
Plant Height: 37–90 cm (1¼–3 ft)
Habitat: Fields and waste ground
Biennial
Flowering Times: May–August

Notes:

The single yellow flower head of this alien species is composed of all ray florets which are surrounded by long pointed green bracts. This flower is among those that open briefly in the morning. We learned this the hard way when after a pleasant brunch, the flower we had expected to photograph had already begun to close up. Another thing that we learned is that several species of goat's beards grow in our area, sometimes right next to each other. They can be told apart by the size relations of the green bracts to the yellow rays. In *T. dubius* the bracts are longer than the rays. Another surprise is that after closing up, this flower metamorphoses into a magnificent "plumose" head with intriguing geometry. Each floret becomes a seed-like fruit connected to a hairy parachute reminiscent of its smaller cousin, the dandelion.

Yellow Goat's Beard

Canada Day / Independence Day

Yellow Goat's Beard

Flower Guide for Holiday Weekends

Plant Name: Honeysuckle, hairy
Lonicera hirsuta Eaton
Flower Size: 2–2.5 cm (³/₄–1 in) long
Plant Height: Vine up to 3 m (9 ft) tall
Habitat: Woodlands
Perennial
Flowering Times: Late June to early August

Notes:

This woody deciduous vine (see photograph below) bears colorful clusters of yellow to orange flowers which arise from the center of a pair of fused leaves. Only the uppermost one or two pairs of leaves are fused. The species name, *hirsuta*, refers to the fact that the undersides and edges of the leaves are hairy. However, the hairs are very small and not easily seen in the photographs.

There are many species of honeysuckle and as their name implies, they are known for their nectar. As kids, many of us extracted a small sweet drop after removing the end of a flower and sucking on the exposed tube. Insects are also attracted to this source of food so flowering honeysuckle vines make good locations for observing or photographing insects.

The orange-red berries which replace the flowers provide food for a variety of birds including grouse and finches, as well as for mammals such as white-tailed deer and snowshoe hares. The berries shown to the right are from a different species, probably *L. tartarica*.

Canada Day / Independence Day

Hairy Honeysuckle

89

Flower Guide for Holiday Weekends

Plant Name: Indian Paintbrush
Castilleja coccinea (L.) Spreng.
Flower Size: 2.5 cm (1 in) but hidden
Plant Height: 30–60 cm (1–2 ft)
Habitat: Meadows, often damp soil
Annual or Biennial
Flowering Times: May–July

Notes:

The scarlet of Indian Paintbrush is visible at great distances in meadows, often to startling effect. The color is actually associated with leaf-like bracts that almost conceal the inconspicuous flowers. The mingling of Indian Paintbrush with buttercups in a wet meadow often looks from a distance like a French impressionist painting. Because the bracts last longer than flowers, paintbrushes appear to "bloom" with successive other species as the season progresses. Hovering insects or birds are needed for pollination as there are no perches available for stationary pollinators. It is a partial root parasite. Vigorous growth of this plant requires both its own photosynthesis and nutrients obtained through a parasitic connection to the roots of another plant. A wide variety of plants serve as hosts. Native Americans used tea made from the flowers for rheumatism and women's ailments as well as a love potion and as a poison!

Canada Day / Independence Day

Indian Paintbrush

Flower Guide for Holiday Weekends

Plant Name: Leek, wild
Allium tricoccum Aiton
Flower Size: 5–7 mm (¼ in), umbel 3.8 cm (1½ in) in diameter
Plant Height: 10–30 cm (4–12 in)
Habitat: Rich moist woods, hardwood forests often in limestone areas
Perennial or Biennial
Flowering Times: June–July
Notes:

One of the curious things about this plant is that the flowers begin to appear only as the leaves are yellowing and dying back. Compare the ground cover in the photograph on the right with that seen on Victoria Day/Memorial Day. It is somewhat disconcerting to see the blooming flowers when the leaves are nowhere to be seen.

This is one of the few flowers to be found within the forest at the beginning of summer. Note that the other flowers in this section are found in more open habitats.

Wild Leek

Canada Day / Independence Day

Wild Leek

Plant Name: Milkweed, common
Asclepias syriaca L.
Flower Size: 1 cm (½ in) wide. Flower head 5–7 cm (2–3 in) in diameter
Plant Height: 60–180 cm (2–6 ft)
Habitat: Old fields, "waste places", roadsides
Perennial
Flowering Times: June–August

Notes:

The common milkweed charms us for many months each year, first with its lovely fragrant flower heads and later with its large pods which open in the autumn to release feathery seeds for dispersal by air currents. The sun glinting off the delicate silk of the seeds stays in our memories and is a favorite photographic subject. The biology of this plant is fascinating. It has a milky sap which oozes out at any point of injury. This fluid contains alkaloids which are bitter and poisonous to most organisms. There are, however, a few insects which have evolved to flourish on milkweed. One is the monarch butterfly whose beautiful black, white and yellow banded caterpillar stores the poisons. Others are the large and small milkweed bugs. Birds are said to learn not to forage on the juvenile or adult members of these species. The adults of both species are easily seen with their bright orange and black coloration. Apparently, other species of black and orange butterflies are protected from predation by their similarity to monarch butterflies. Milkweed reproductive biology is also interesting. The 75 or so flowers in each flower head attract a variety of insects to their nectar in exchange for distributing pollen for fertilization. Our photograph shows numerous moths on the flower heads. Despite the expense of producing all the flowers, generally, only two pods will develop from a flower head. Milkweeds use the airborne seeds for dispersal and underground rhizomes to expand the population of plants once a plant has become established.

Milkweed silk historically has been used for pillow and mattress stuffing, candle wicks and amazingly for life vests, as it is, weight for weight, five times more buoyant than cork and one sixth as heavy as wool while providing greater insulation!

Canada Day / Independence Day

Common Milkweed

Flower Guide for Holiday Weekends

Plant Name: Raspberry, purple-flowering
   Rubus odoratus L.
Flower Size: 2.5–5 cm (1–2 in) diameter
Plant Height: 90–180 cm (3–6 ft)
Habitat: Rocky woods and thickets. Moist shaded forest edges and roadsides
Perennial
Flowering Times: June–September

Notes:

This thornless perennial shrub bears beautiful purplish pink flowers which despite the Latin name *odoratus* have only a subtle fragrance. The large palmate leaves (reminiscent of maple leaves) form a suitable backdrop to a profusion of flowers. As these plants grow in moist, shady but open areas and as they lack the prickles associated with so many of their raspberry cousins, we can approach them without trepidation. The pastel flower is replaced by a rather large red berry which unfortunately is neither juicy nor tasty to humans although wild life seem to like them.

The plant has many uses. Not only the fruit (see below) but also the buds and leaves are eaten by birds and mammals including people. The leaves are very astringent. The Cherokee made a decoction from them which they used as a wash for ulcers and sores, and as a remedy for menstrual and bowel complaints. A somewhat more unusual use was made of the leaves by *coureurs de bois* (17th Century unlicenced fur traders in what is now Quebec, Canada) who put them in their leather shoes to protect their feet.

Purple-Flowering Raspberry

Canada Day / Independence Day

Purple-Flowering Raspberry

Smooth Rose

Plant Name: Rose, smooth
Rosa blanda Aiton
Flower Size: 5–6 cm (2–2½ in) wide
Plant Height: Shrub up to 1.5 m (4½ ft) tall
Habitat: Clearings, fields, roadsides, open forests. Moist to dry soil
Perennial
Flowering Times: May to early July

Notes:

Of all the many species of wild roses this is the most easily identified. Also known as the meadow rose, it is distinguished by its smooth stems, devoid of prickles and thorns, although vigorous shoots may be prickly near their base. This may well be the species inspiring the American composer, Edward MacDowell to write the well known song, "To a Wild Rose".

The fruits, the familiar red rose hips, smooth and round to egg-shaped, 1–1.5 cm (⅜–⅝ in) in diameter, are found from August to early October. They are eaten by aboriginal peoples, black bears, snowshoe hares, grouse and pheasants. Rose hip tea is still a popular drink.

The inner bark of the roots was used in a medicine for cataracts.

Canada Day / Independence Day

Smooth Rose

Flower Guide for Holiday Weekends

Striped Coral Root

Plant Name: Striped coral root
Corallorhiza striata Lindl.
Flower Size: 8–20 mm (³/₈–³/₄ in) long. Raceme with 3–30 flowers
Plant Height: 20–50 cm (8–20 in)
Habitat: Rich moist woods with calcareous or alkaline soil
Perennial
Flowering Times: May–August

Notes:

This strange plant, without a sign of green, is an orchid which lacks chlorophyll. Unlike green plants which can make sugars using photosynthesis, the striped coral root must gain its nutrition by other means. On the basis of what is known for the spotted coral root (*C. maculata*) it is thought that the striped coral root obtains carbon compounds through association with fungi (mycorrhizae). It is hard to imagine that it is fairly closely related to the Lady's Slippers. The name coral root comes from the Greek words for coral and root (*rhiza*), but the coral-like "roots" are in fact rhizomes. The pinkish-yellow or white flowers with conspicuous reddish purple stripes also conjure up visions of coral. Considering that most plants were found useful by aboriginal peoples, it is not surprising to learn that the root was used as a sedative and to reduce fevers.

Canada Day / Independence Day

Striped Coral Root

Flower Guide for Holiday Weekends

Plant Name: Sweet Clover, white
Melilotus alba Desv. Medik.
Flower Size: 6 mm (¼ in) long borne in long spike-like clusters up to 20 cm (8 in) long
Plant Height: 90–240 cm (3–8 ft)
Habitat: Roadsides and fields
Biennial
Flowering Times: May–October
Notes:

This species is taller than most other clovers, but it has the typical clover tripartite leaves.

Known as the honey plant because its nectar is the chief source of clover honey, the crushed flowers have the sweet fragrance of new mown hay due to the coumarin they contain. Introduced from Europe, it is planted as a pasture crop for nitrogen enrichment of the soil. Its roots bear visible nodules containing bacteria, which can take nitrogen from the air and transform it into a biologically useful form.

Moldy sweet clover hay or silage causes severe hemorrhagic disease in cattle. The molds convert the coumarin in the clover to a hydroxy-coumarin (dicoumarol) which like warfarin, another hydroxycoumarin, is an anti-clotting agent.

Canada Day / Independence Day

White Sweet Clover

Civic Holiday/Labor Day

Baneberry, red — *Acetaea rubra* (Aiton) Willd. 106
Bergamot — *Monarda fistulosa* L. 108
Blue vervain — *Verbena hastata* L. 110
Boneset — *Eupatorium perfoliatum* L. 112
Cardinal flower — *Lobelia cardinalis* L. 114
Cohosh blue — *Caulophyllum thalictroides* (L.) Michx. 116
Evening primrose — *Oenothera biennis* L. 118
False Solomon's seal, star-flowered —
   Maianthemum stellatum (L.) Link 120
Goat's beard, yellow — *Tragopogon dubius* Scop. 122
Goldenrod — *Solidago* L. spp. 124
Helleborine — *Epipactis helleborine* (L.) Crantz. 126
Leek, wild — *Allium tricoccum* Aiton 128
Milkweed, common — *Asclepias syrica* L. 130
Mint, wild — *Mentha arvensis* L. subsp. *borealis*
   (Michx.) R.L. Taylor & Macbryde 132
Ragweed — *Ambrosia artemisiifolia* L. 134
Raspberry, purple-flowering — *Rubus odoratus* L. 136
Staghorn sumac — *Rhus typhina* L. 138
Teasel — *Dipsacus fullonum* L. subsp. *sylvestris*
   (Hudson.) Clapham 140
Thistle, bull — *Cirsium vulgare* (Savi) Ten. 142

Red Baneberry

Plant Name: Baneberry, red
  Actaea rubra (Aiton) Willd.
Flower Size: Cluster diameter is 2.5–3.75 cm (1–1½ in)
Plant Height: 30–80 cm (1 ft–2 ft 8 in)
Habitat: A variety of woodlands
Perennial
Flowering Times: May–June

Notes:

The baneberries look like small bushes and are found scattered in woods. The most striking aspect of the red baneberry is the shiny red cluster of berries which appear in late summer. A shaft of sunlight coming through the forest canopy and reflecting off the incredibly shiny berries enlivens many a late summer walk. White baneberry grows in similar areas and is distinguished by its white berries (called 'doll's eyes') on thick flower stalks.

Despite the fact that baneberries have toxic substances in all parts of the plant, aboriginal women apparently used a tea prepared from roots after giving birth. The tea was also drunk as a cure for syphilis and rheumatism.

Ruffed grouse eat the berries with impunity.

Civic Holiday / Labor Day

Red Baneberry

Plant Name: Bergamot
Monarda fistulosa L.
Flower Size: 2.5 cm (1 in) long in clusters 2.5–4 cm (1–1½ in) across
Plant Height: 60–120 cm (2–3 ft)
Habitat: Dry fields and sunny thickets particularly, edges, and high calcium areas
Perennial
Flowering Times: July–September
Notes:

As the spring flower season draws to a close we find consolation in looking forward to the appearance of the pink/lavender flowers of bergamot. Thriving in sunny, open areas they eventually form clumps of stems emanating from the same root stock. In addition to their beauty, like many other mints, they are aromatic; if you crush parts of the plant (including the remains of the flower heads, in winter) you will find a distinctly Earl Grey tea smell on your fingers. Indeed, the oil of bergamot used in that tea comes from a close relative.

 Traditionally, the leaves or blossoms of the wild plant have been used to flavor tea and other dishes and the Ojibwa are said to have used the roots to relieve gastric distress. A modern herb encyclopedia mentions that the leaves are used to flavor meat, beans and in teas. In addition they have been used to benefit digestion, cure colds, sore throat, headaches, fevers, and skin eruptions!

Civic Holiday / Labor Day

Bergamot

Plant Name: Blue Vervain
Verbena hastata L.
Flower Size: 3 mm wide (1/8 in)
Plant Height: 60–180 cm (2–6 ft)
Habitat: Damp thickets, often near river and lake shores, roadsides
Perennial
Flowering Times: July–September
Notes:

Blue vervain is a tall plant found in moist habitats in the company of plants like joe-pye weed and boneset. Unlike many plants with clustered or stalked flowers, the vervain has only a few flowers in bloom at any given time. This results in months of flowering and a reliable plant for photographing nectar seeking insects. During the winter, the stalks of the vervain often remain erect and may be used as an interesting addition to a dried winter bouquet.

Hummingbirds and butterflies find it attractive. It is pollinated by bumblebees.

Ancient Celtic, Druid, German and Roman cultures held a European relative in high regard or as a sacred plant.

The seeds are eaten by birds and rabbits, and by humans who roast and grind them into a flour.

Native Americans and 19th century physicians used the plants in remedies for coughs, colds, stomach cramps and bowel complaints. It is also said to be an aphrodisiac.

Civic Holiday / Labor Day

Blue Vervain

Plant Name: Boneset
Eupatorium perfoliatum L.
Flower Size: Individual flowers 6 mm (¼ in) across in flat clusters 10–16 mm (³/₈–⁵/₈ in) in diameter
Plant Height: 60–150 cm (2–5 ft)
Habitat: Moist meadows and woods
Perennial
Flowering Times: July–October

Notes:

In wet areas and along streams, tall sturdy plants begin to flower as summer wears on. Boneset is one of these. Its leaves are distinctive in that they surround the stem which may have given rise to the idea that the plant was good for setting bones and hence its popular name. It was not a plant highly regarded by children of early settlers who were forced to drink nauseating "boneset tea" if they fell ill with fever. We, however, find the natural "gardens" it forms with Joe-pye weed (a close relative) and cardinal flowers, delightful.

Civic Holiday / Labor Day

Boneset

Plant Name: Cardinal flower
Lobelia cardinalis L.
Flower Size: 3.8 cm (1½ in)
Plant Height: 60–120 cm (2–4 ft)
Habitat: Damp sites, especially along waterways, moist rich soil. Often found in association with boneset and blue vervain
Perennial
Flowering Times: July–September

Notes:

Everything about the Cardinal flower delights. The crimson color is magnificent, the structure of the flower both intricate and beautiful and the biology fascinating. The plant is most often seen along waterways, often singly, but in some places they are more numerous along a stream bank creating a stunning view. Where boneset and cardinal flowers grow side by side the contrasting white and red blooms provide a festive display for several weeks (see habitat photograph).

They propagate using rhizomes as well as seeds. Pollination in this plant is interesting because it depends on hummingbirds and certain moths using their long mouth parts to get nectar from the bottom of the long upward pointing flower tube and thereby inadvertently procuring and carrying pollen from plant to plant. There is even more to the story. Cardinal flowers mature first as males (see close-up photograph with yellow pollen) and later the pollen bearing area is replaced by a female structure for receiving pollen. This results in male flowers being at the top of the stalk and the females at the bottom. At times hummingbirds feed first from the bottom of the flower spike and then work their way up the stalk to the younger flowers. This tends to ensure that pollen from the top of a plant is deposited on the female flowers of the next plant they visit.

The Cherokees used the roots in a preparation against syphilis and worms. It was also an ingredient of love potions.

Civic Holiday / Labor Day

Cardinal Flower

Plant Name: Cohosh, blue
Caulophyllum thalictroides (L.) Michx.
Flower Size: 3–6 cm (1¼–2½ in)
Plant Height: 30–80 cm (1 ft–2 ft 8 in)
Habitat: Moist woods and hardwood forests, particularly in limestone areas
Perennial
Flowering Times: April–May

Notes:

The large dark blue berry-like seeds (see photograph below) are more vivid than the flowers (lower right). Cross-fertilization is ensured by the female and male parts of the flower maturing at different times.

When roasted and ground the seeds have been used as a coffee substitute.

The roots were used by Native Americans to facilitate childbirth and it was listed in the US Pharmacopoeia until 1905. Latter day herbalists still make medicinal claims for treating everything from worms to gout but admit that patients with hypertension or heart disease should give this curative a miss.

Civic Holiday / Labor Day

Blue Cohosh

Flower Guide for Holiday Weekends

Plant Name: Evening primrose
Oenothera biennis L.
Flower Size: 2–5 cm (³/₄–2 in)
Plant Height: 30–150 cm (1–5 ft)
Habitat: Clearings, dry fields, roadsides, and waste areas
Biennial
Flowering Times: June–September

Notes:

Evening primrose is a plant of disturbed ground in open areas. It is very hardy and is one of the relatively few North American plants which have settled successfully in Europe. Among its interesting features is its tendency to bloom in the late afternoon and evening when it will be pollinated by insects of the night. Another unusual feature is that the plant first produces a rosette of leaves and if this rosette does not grow big enough to support flowering the first year, it will over-winter before flowering. The characteristic woody winter stalks protrude from the snow, reminding us of summer delights. to come. The primroses have been extensively studied by cytogeneticists because in some races, chromosomes form circles during cell divisions producing the ovules and pollen. This feature should reduce chromosome recombination which is believed to be important for adaptation to changing environments. Not having listened to scientists, they are happily widespread, numerous and adaptable.

The thick taproot in its first year when it is tender, is nutritious and tasty when boiled and new leaves are used in salad. Goldfinches and other birds like the seeds which are also eaten by people.

Oil from seeds has been found to be rich in an essential fatty acid considered by some to be helpful in treating everything from multiple sclerosis and menstrual cramps to hyperactivity in children.

Civic Holiday / Labor Day

Evening Primrose

Plant Name: False Solomon's seal, star-flowered
Maianthemum stellatum (L.) Link
(formerly *Smilacina stellata* Desf.)
Flower Size: 6 mm (¼ in) long
Plant Height: 20–50 cm (8–20 in)
Habitat: Prefers moist soil and is found along roadsides and in damp woods, but also grows on sand dunes
Perennial
Flowering Times: May–June

Notes:

This member of the lily family has four or more alternate narrow leaves clasping the stem.

The delicate creamy-white flowers at the ends of stems seen in the spring give rise to clusters of green berries which become attractively candy-striped before turning dark red (see insert) and eventually black.

Under the right conditions, dense patches of these plants are found along roadsides as shown in the large photograph.

Civic Holiday / Labor Day

Star-Flowered False Solomon's Seal

Yellow Goat's Beard

Plant Name: Goat's beard, yellow
Tragopogon dubius Scop.
Flower Size: 2.5–6.3 cm (1–2½ in)
Plant Height: 37–90 cm (1¼–3 ft)
Habitat: Fields and waste ground
Biennial
Flowering Times: May–August

Notes:

The single yellow flower head of this alien species is composed of all ray florets which are surrounded by long pointed green bracts. This flower is among those that open briefly in the morning. We learned this the hard way when after a pleasant brunch, the flower we had expected to photograph had already begun to close up. Another thing that we learned is that several species of goat's beards grow in our area, sometimes right next to each other. They can be told apart by the size relations of the green bracts to the yellow rays. In *T. dubius* the bracts are longer than the rays. Another surprise is that after closing up, this flower metamorphoses into a magnificent "plumose" head with intriguing geometry. Each floret becomes a seed-like fruit connected to a hairy parachute reminiscent of its smaller cousin, the dandelion.

Civic Holiday / Labor Day

Yellow Goat's Beard

123

Flower Guide for Holiday Weekends

Plant Name: Goldenrod
   Solidago L. spp.
Flower Size: 3 mm (⅛ in) diameter
Plant Height: 30–150 cm (1–5 ft)
Habitat: Roadsides, disturbed ground, old fields
Perennial
Flowering Times: Late July to early October. *S. altissima* into December.

Notes:

Goldenrods are blamed for allergies but their pollen is large and sticky and dispersed by insects not wind. Ragweed, the real culprit, blooms at the same time but its flowers are inconspicuous. The two are shown together in the lower left photograph, the greenish spikes below the goldenrod are ragweed inflorescences.

Solidago means "to make whole" and originates from the traditional medicinal uses of some species. Not only are the numerous species difficult to identify but there is extensive hybridization between them. Distinguishing features are leaf size and position, inflorescence shape and position as well as minute differences in the individual flowers. The goldenrods shown in the two photographs here are different species. Note that the leaves are obviously different.

Insect induced galls are often found on the stems of various goldenrod species. Leaves, flowers and seeds are eaten by a wide variety of mammals and birds.

Aboriginal people chewed crushed flowers of *S. canadensis* for sore throats. Flower tea was used to treat fevers and snake bites. A yellow dye was prepared from the flowers.

S. canadensis contains quercetin which is used to treat hemorrhagic nephritis.

Civic Holiday / Labor Day

Goldenrod

Plant Name: Helleborine
    Epipactis helleborine (L.) Crantz
Flower Size: Lower petal 1–1.5 cm ($^3/_8$–$^5/_8$ in) long, 4–8 mm ($^1/_6$–$^1/_3$ in) wide, lateral petals and sepals 10–14 mm ($^3/_8$–$^1/_2$ in) long
Plant Height: 30–80 cm (1–3 ft)
Habitat: Variety of soils and degrees of moistness, deciduous forests and forest edges, and disturbed areas Oak, maple woods
Perennial
Flowering Times: July–August

Notes:

Helleborine flowers are a mid-summer treat. This relatively common orchid has subtly colored flowers as well as an interesting shape. Because the flowers are vertically arranged and relatively small, some effort is required to position oneself to appreciate their beauty. The lower buds on the stalk mature first and may become seed capsules while the upper flowers are still in bloom. They are pollinated by wasps.

In Europe it was used for medicinal purposes which may explain its introduction to North America where it is very successful.

Civic Holiday / Labor Day

Helleborine

Plant Name: Leek, wild
  Allium tricoccum Aiton
Flower Size: 5–7 mm (¼ in), umbel 3.8 cm (1½ in) in diameter
Plant Height: 10–30 cm (4–12 in)
Habitat: Rich moist woods, hardwood forests often in limestone areas
Perennial or Biennial
Flowering Times: June–July

Notes:

One of the curious things about this plant is that the flowers begin to appear only as the leaves are yellowing and dying back. It is somewhat disconcerting to see blooming flowers when the leaves are nowhere to be seen.

Green, then black shiny seed heads are formed after the leek has flowered. These heads atop their long stem may persist through the winter and be seen among the lush green leaves the following spring. As may be seen in the photographs on the right the seed heads mirror the morphology of the flower heads.

Wild Leek

Civic Holiday / Labor Day

Wild Leek

129

Plant Name: Milkweed, common
Asclepias syriaca L.
Flower Size: 1 cm (½ in) wide. Flower head 5–7 cm (2–3 in) in diameter
Plant Height: 60–180 cm (2–6 ft)
Habitat: Old fields, "waste places", roadsides
Perennial
Flowering Times: June–August

Notes:

The common milkweed charms us for many months each year, first with its lovely fragrant flower heads and later with its large pods which open in the autumn to release feathery seeds for dispersal by air currents. The sun glinting off the delicate silk of the seeds stays in our memories and is a favorite photographic subject. The biology of this plant is fascinating. It has a milky sap which oozes out at any point of injury. This fluid contains alkaloids which are bitter and poisonous to most organisms. There are, however, a few insects which have evolved to flourish on milkweed. One is the monarch butterfly whose beautiful black, white and yellow banded caterpillar stores the poisons. Others are the large and small milkweed bugs. Birds are said to learn not to forage on the juvenile or adult members of these species. The adults of both species are easily seen with their bright orange and black coloration. Apparently, other species of black and orange butterflies are protected from predation by their similarity to monarch butterflies. As monarch butterflies are gathering for their southward migration and students are once again in classes, the green pods of the milkweed dry into tawny colored pods which eventuallysplit open to allow the wind to help their beautifully packaged seeds to become airborne (top photograph). Milkweeds use the airborne seeds for dispersal and underground rhizomes to expand the population of plants once a plant has become established.

Milkweed silk historically has been used for pillow and mattress stuffing, candle wicks and amazingly for life vests, as it is, weight for weight, five times more buoyant than cork and one sixth as heavy as wool while providing greater insulation!

Civic Holiday / Labor Day

Common Milkweed

Plant Name: Mint, wild
Mentha arvensis L. subsp. *borealis* (Michx.) R.L. Taylor & Macbryde
Flower Size: 3.2 mm (⅛ in) wide; 6 mm (¼ in) long
Plant Height: 30–60 cm (1–2 ft) tall
Habitat: Damp and wet places
Perennial
Flowering Times: July–September

Notes:

This native perennial is named after the Greek nymph Menthe, who was changed into a plant by a jealous queen of the underworld, Proserpine.

The pale lilac or white flowers grow in clusters around the square stem where the paired leaves arise. The plant is often difficult to see when it is growing in crowded conditions but its smell as it gets crushed underfoot is unmistakable.

The highly aromatic leaves are used in sauces, jellies, and beverages. Native Americans made a leaf tea which they drank as a remedy for colds, sore throats, fevers, indigestion, diarrhea, and headaches.

Civic Holiday / Labor Day

Wild Mint

Plant Name: Ragweed
Ambrosia artemisiifolia L.
Flower Size: Monoecious. Male flowers small yellowish-green in heads containing florets in clusters 2.5–15 cm (1–6 in) long. Female flowers small green and stalkless at leaf axils
Plant Height: 30–150 cm (1–5 ft)
Habitat: Roadsides, waste ground, cultivated fields
Annual
Flowering Times: July–October

Notes:

Copious wind borne pollen from the hundreds of male flowers (the flowers with yellow pollen in inset) on each plant causes untold misery to allergy sufferers. A female flower with protruding styles is also shown in the inset. Contrary to popular misconception it is ragweed not goldenrod, the conspicuous yellow flower in the large photograph, which is the culprit. Brightly colored flowers like goldenrod depend on insects for pollination and produce fewer pollen grains than wind dependent species. Few things annoy botanists more than advertisements for anti-allergy medications featuring a flower obviously pollinated by insects.

The name *Ambrosia* meaning "food for the gods that imparts immortality", seems unusually inappropriate.

Native Americans took teas prepared from leaves and roots for a range of ailments from fever to stroke. The leaves were rubbed on insect bites and infected skin injuries.

Civic Holiday / Labor Day

Ragweed

Flower Guide for Holiday Weekends

Plant Name: Raspberry, purple-flowering
Rubus odoratus L.
Flower Size: 2.5–5 cm (1–2 in) diameter
Plant Height: 90–180 cm (3–6 ft)
Habitat: Rocky woods and thickets. Moist shaded forest edges and roadsides
Perennial
Flowering Times: June–September

Notes:

This thornless perennial shrub bears beautiful purplish pink flowers which despite the Latin name *odoratus* have only a subtle fragrance. The large palmate leaves (reminiscent of maple leaves) form a suitable backdrop to a profusion of flowers. As these plants grow in moist, shady but open areas and as they lack the prickles associated with so many of their raspberry cousins, we can approach them without trepidation. The pastel flower is replaced by a rather large red berry (see photograph below) which unfortunately is neither juicy nor tasty to humans although wild life seem to like them.

The plant has many uses. Not only the fruit but also the buds and leaves are eaten by birds and mammals including people. The leaves are very astringent. The Cherokee made a decoction from them which they used as a wash for ulcers and sores, and as a remedy for menstrual and bowel complaints. A somewhat more unusual use was made of the leaves by *coureurs de bois* (17th Century unlicensed fur traders in what is now Quebec, Canada) who put them in their leather shoes to protect their feet.

Civic Holiday / Labor Day

Purple-Flowering Raspberry

Plant Name: Staghorn Sumac
Rhus typhina L.
Flower Size: Both male and female flowers are 3 mm (1/8 in) broad
Plant Height: Shrubs or small trees up to 6 m (18 ft)
Habitat: Open fields and roadsides. Forms thickets in well drained soils.
Perennial
Flowering Times: July into September

Notes:

In winter the branches stick up in a pattern resembling the antlers of deer, hence staghorn. This sumac is distinguished by the hairiness of its twigs and leaf stalks. The soft down makes the under surfaces of the leaves appear almost white. Although this sumac is a common sight along roads it is not generally realized that male and female flowers are borne on separate plants. Both are inconspicuous greenish yellow and occur in terminal erect cone-shaped inflorescences. The male flowers (see inset on right) are distinguished from the females by having several prominent anthers with yellow pollen. With the development of the fruits the female inflorescences become the familiar red cones. The color derives from the dense covering of crimson hairs on the fruits.

Many people wonder whether staghorn sumac is poisonous. It is not. Poison sumac is a different species and is also known as swamp sumac reflecting its very different habitat.

The twigs and leaves are browsed by deer, moose and rabbits. The fruits were used in traditional remedies for coughs. Teas prepared from leaves, fruits, roots and bark were used for a variety of ailments from sore throats to worms. A recent book on wild foods (Stalking the Wild Asparagus by Euell Gibbons) describes how to make a sumac drink using a washing machine! Another contemporary use of the red fruits is as a combustible material in bee smokers.

Civic Holiday / Labor Day

Staghorn Sumac

Plant Name: Teasel
Dipsacus fullonum L. subsp. *sylvestris* (Hudson) Clapham
(formerly *D. sylvestris* Hudson)

Flower Size: Individual flowers 1.3 cm (½ in) long in heads 3.8–10 cm (1½–4 in) long and 2.5–5 cm (1–2 in) wide.

Plant Height: 60–180 cm (2–6 ft)

Habitat: Roadsides, waste places, and old fields

Biennial

Flowering Times: July–October

Notes:

Flowers, of this European import, first open in a belt around the middle of the head. New flowers open daily in both directions so that in time there are two bands of flowers. The heads are at first egg-shaped then become cylindrical. The stems, flower stalks and the midribs of the lanceolate leaves are prickly and there are spiny bracts between the flowers.

When dried, the flower heads were used to raise the nap, or tease, cloth. It was therefore natural for colonists to bring this useful plant with them to North America.

Civic Holiday / Labor Day

Teasel

Plant Name: Thistle, bull
Cirsium vulgare (Savi) Ten.
Flower Size: Heads 3.8–5 cm (½–2 in) wide
Plant Height: 60–180 cm (2–6 ft)
Habitat: Roadsides, pastures, and waste places
Biennial
Flowering Times: June–September

Notes:

A very prickly plant, the spiniest of the thistles, even the flower heads are surrounded by spiny bracts. While attractive to observe, close proximity can be painful. At such times there is consolation in the knowledge that goldfinches use thistle down to line their nests.

A related thistle, Canada thistle, was imported from Europe to Canada and has spread to the United States. Because its rootstock is below cultivation depth it can spread under ground and annihilate crops in farmers' fields. Ironically, this pest has smooth stems.

Civic Holiday / Labor Day

Bull Thistle

Glossary

Alkaloid	A group of compounds found in plants which may have medicinal or toxic effects.
Alternate	Branches or leaves which emerge from the stem at different heights. Cf. opposite
***Anther**	Pollen bearing part of a stamen
Biennial	A plant which lives for two years flowering only in the second year.
Bract	A modified leaf sometimes found below sepals.
Chlorophyll	The green pigments in plants important for photosynthesis
Chromosome	Structures within each cell carrying hereditary material
Corm	Enlarged base of a stem which both stores food and is a means of vegetative reproduction
Deciduous	Woody plants whose leaves are shed annually
Disk florets	Small tubular flowers found in compound flower heads in the daisy family
Emetic	Induces vomiting
Fertilization	Sexual reproduction in plants, the union of pollen with ovule leading to seed production
Floret	An individual flower in the compound flower of the daisy family
Glucoside	A compound of glucose and other substances
Glycoside	A compound of a sugar with other substances
Inflorescence	A shoot bearing a number of flowers
Lanceolate	Slender tapered shape
Opposite	Branches or leaves emerge from the stem at the same height on opposite sides
***Ovary**	Structure enclosing ovules
***Ovule**	A female part of a plant which unites with pollen in sexual reproduction
Palmate	Palm-like shape

*Petal	One of the modified leaves in the inner whorl of flowers, often brightly colored and conspicuous
*Pistil	Female portion of a plant
Pollen	Male part of plant which unites with an ovule to produce a seed
Pollination	Transfer of pollen to the female part of the plant, typically by wind or insects or birds
Ray florets	Flattened ray-shaped flowers found in the compound flower heads of the daisy family
Rhizome	Underground stem, usually horizontal, which is a means of vegetative reproduction and may also store food
*Sepal	Outermost parts of flowers which protect the bud. Although usually they remain green they may become colored, e.g., in anemone and iris
*Stamen	Male part of flower
*Style	The stalk connecting the stigma, the area at the tip of the style receptive to pollen, with the ovary
Taproot	Main large root growing downward
Vegetative reproduction	Reproduction from parts of a plant not involving fertilization

*See diagram of floral parts

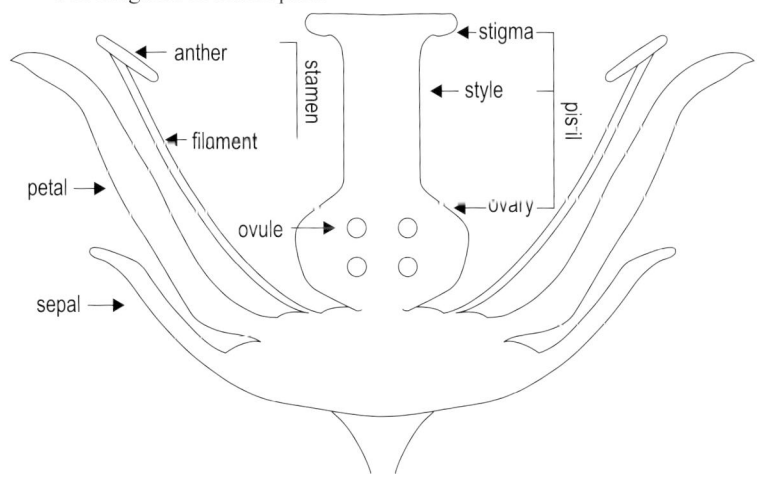

Labels: anther, stamen, stigma, style, pistil, filament, petal, ovule, ovary, sepal

Bibliography

Bown, D. 1995. *The Royal Horticultural Society Encyclopedia of Herbs.* Dorling Kindersley, London.

Britton, N.L., and Brown, A. 1970. *An Illustrated Flora of the Northern United States and Canada (Volumes 1–3).* Dover, New York.

Bruce-Grey Plant Committee 1997 *The Orchids of Bruce and Grey.* Owen Sound Field Naturalists

Bruce-Grey Plant Committee 2000 *The Asters, Goldenrods and Fleabanes of Grey and Bruce Counties.* Owen Sound Field Naturalists.

Chambers, B., Legasy, R., and Bentley, C.V. 1996. *Forest Plants of Central Ontario.* Lone Pine, Edmonton.

*Coffey, T. 1993. *The History and Folklore of North American Wildflowers.* Houghton Mifflin, Boston New York.

Collins, M. 2000 The influence of host-plant species and gall diameter on the distribution of three goldenrod gall-makers; their predators and parasitoids. Zoology Department, University of Toronto, Toronto.

Courtenay, B., and Zimmerman, J.H. 1972. *Wildflowers and Weeds.* Van Nostrand Reinhold Ltd.

*Densmore, F. 1974. *How Indians Use Wild Plants for Food, Medicine and Crafts.* Dover, New York.

Farrar, J.L. 1995. *Trees in Canada.* Fitzhenry & Whiteside Ltd and Canadian Forest Service, Markham, Ontario

*Foster, S., and Duke, J.A. 1990. *A Field Guide to Medicinal Plants. Eastern and Central North America.* Houghton Mifflin, Boston New York

Kershaw, L. 2001. *Trees of Ontario.* Lone Pine, Edmonton.

Morton, J.K. and Venn, J.M. 1990. *A Checklist of the Flora of Ontario — Vascular Plants.* University of Waterloo Biology Series 34.

Newmaster, S.G., Lehla, A., Uhlig, P.W.C., McMurray, S., and Oldham, M.J. 1998. *Ontario Plant List.* Forest Research Paper No. 123. Sault Ste. Marie ON, Ontario Forest Research Institute.

Newmaster, S.G., Harris, A.G., and Kershaw, L.J. 1997. *Wetland Plants of Ontario.* Lone Pine, Edmonton.

Bibliography

Newcomb, L. 1977 *Newcomb's Wildflower Guide*. Little Brown & Company Boston

Niering, W.A. 1985. *Wetlands*. National Audubon Society Nature Guides.

Niering, W.A., and Olmstead, N.C. 1979. *The Audubon Society Field Guide to North American Wildflowers. Eastern Region*. Knopf, New York.

Peterson, L.A,. 1977. *A Field Guide to Edible Wild Plants Eastern and Central North America*. Houghton Mifflin, Boston New York.

Peterson, R.T and McKenney, M. 1968. *A Field Guide to Wildflowers. Northeastern and Northcentral North America*. Houghton Mifflin, Boston New York

Royer, F., and Dickenson, R. 1999. *Weeds of Canada and the Northern United States*. Lone Pine, Edmonton.

Scoggan, H.J. 1978–1979. *The Flora of Canada*. National Museum of Canada, Ottawa.

Semple, J.C., and Ringuis, G.S. 1983. *The Goldenrods of Ontario. Solidago and Euthamia*. Department of Biology, University of Waterloo, Waterloo, Ontario.

Shinkel, D., and Mohrhardt, D. 1994. *Favorite Wildflowers of the Great Lakes and Northeastern U.S.* Thunder Bay Press, Thunder Bay.

Soper, J.H., and Heimburger, M.L. 1982. *Shrubs of Ontario*. Royal Ontario Museum, Toronto.

Stokes, D., and Stokes, L. 1986. *Stokes Field Guide to Enjoying Wildflowers*. Little, Brown, and Co., Boston.

*Swerdlow, J.L. 2000. *Nature's Medicine. Plants That Heal*. National Geographic Society, Washington D.C.

Symmonds, G.D. 1963. *The Shrub Identification Book*, George J. McLeod Ltd.

Venning, F.D. 1984 *A Guide to Field Identification. Flowers of North America*. Golden Press, New York.

Zichmanis, Z., and Hodgins, J. 1982. *Flowers of the Wild. Ontario and the Great Lakes Region*. Oxford University Press, Toronto.

*Indicates sources of further information regarding uses.

Flower Guide for Holiday Weekends

Common Name Index

A
Anemone, Canada   74, 75

B
Baneberry, red   12, 13, 106, 107
Beardtongue, hairy   14, 15, 76, 77
Bellwort, large-flowered   16, 17, 78, 79
Bergamot   2, 108, 109
Bloodroot   18, 19
Blue cohosh   22, 23, 116, 117
Blue-eyed grass   20, 21, 36, 80, 81
Blue vervain   110, 111, 114
Boneset   110, 112–114

C
Cardinal flower   112, 114, 115
Cohosh, blue   22, 23, 116, 117
Columbine   24, 25, 52, 82, 83

D
Dogbane, spreading   84, 85

E
Evening primrose   118, 119

F
False Solomon's seal,
    star-flowered   26, 27, 120, 121

G
Ginger, wild   28, 29
Goat's beard, yellow   20, 80, 86, 87, 122, 123
Goldenrod   124, 125, 134

H
Hawthorn   30, 31
Helleborine   126, 127
Hepatica   32, 33
Honeysuckle, hairy   88, 89

I
Indian paintbrush   34, 35, 90, 91
Iris, blue flag   36, 37

J
Juneberry   38, 39

L
Lady's slipper, large yellow   40, 41, 62, 100
Leek, wild   42, 43, 92, 93, 128, 129

M
Marsh marigold   44, 45
Meadow rue, early, ♂ & ♀   46, 47
Milkweed, common   94, 95, 130, 131
Mint, wild   132, 133

P
Pin cherry   48, 49
Polygala, fringed   50, 51
Prairie smoke   24, 52, 53, 82
Pussytoes, smaller   54, 55

R
Ragweed   124, 134, 135
Raspberry, purple-flowering   96, 97, 136, 137
Rose, smooth   98, 98

S
Spring beauty   56, 57
Staghorn sumac   26, 27, 120, 121
Solomon's seal, star-flowered
    (see False Solomon's seal)   26, 27, 120, 121
Strawberry, barren   58, 59
Strawberry, common   60, 61
Striped coral root   62, 63, 100, 101
Sweet clover, white   102, 103

T
Teasel   140, 141
Thistle, bull   142, 143
Trillium, large-flowered   64, 65
Trillium, purple   66, 67
Trout lily   20, 68, 69, 80

V
Violet, downy yellow   70, 71

Scientific Name Index

A
Actaea rubra (Aiton) Willd.   12, 106
Allium tricoccum Aiton   42, 92, 128
Ambrosia artemisiifolia L.   134
Amelanchier Medik. spp.   38
Anemone acutiloba (DC.) G. Lawson 32
Anemone canadensis L.   74
Antennaria neodioica Greene *howellii* Greene subsp. *neodioica* (Greene) R.J. Bayer   54
Apocynum androsaemifolium L.   84
Aquilegia canadensis L.   24, 82
Asarum canadense L.   28
Asclepias syriaca L.   94, 130

C
Caltha palustris L.   44
Castilleja coccinea (L.) Spreng.   34, 90
Caulophyllum thalictroides (L.) Michx.   22, 116
Cirsium vulgare (Savi) Ten.   142
Claytonia virginica L.   56
Corallorhiza striata Lindl.   63, 100
Crataegus L. spp.   30
Cypripedium calceolus L. var. *pubescens* (Willd.) Correll   40

D
Dipsacus fullonum L. subsp. *sylvestris* (Hudson.) Clapham   140

E
Epipactis helleborine (L.) Crantz   126
Erythronium americanum Ker Gawl. subsp. *americanum*   68
Eupatorium perfoliatum L.   112

F
Fragaria virginiana Miller subsp. *virginiana*   60

G
Geum triflorum Pursh   52

I
Iris versicolor L.   36

L
Lobelia cardinalis L.   114
Lonicera hirsuta Eaton   88

M
Maianthemum stellatum (L.) Link   26, 120
Melilotus alba Desv. Medik.   102
Mentha arvensis L. subsp. *borealis* (Michx.) R.L. Taylor & Macbryde   132
Monarda fistulosa L.   108

O
Oenothera biennis L.   118

P
Penstemon hirsutus (L.) Willd.   14, 76
Polygala paucifolia Willd.   50
Prunus pennsylvanica L. f.   48

R
Rhus typhina L.   138
Rosa blanda Aiton   98
Rubus odoratus L.   96, 136

S
Sanguinaria canadensis L.   18
Sisyrinchium montanum Greene   20, 80
Solidago L. spp.   124

T
Thalictrum dioicum L.   46
Tragopogon dubius Scop.   86, 122
Trillium erectum L.   66
Trillium grandiflorum (Michx.) Salisb.   64

U
Uvularia grandiflora Sm.   16, 78

V
Verbena hastata L.   110
Viola pubescens Aiton   70

W
Waldsteinia fragarioides (Michx.) Tratt.   58